BEYOND THE SUN

Discovering the Secrets of Our Solar System

Philipp Frühwirth

CONTENTS

INTRODUCTION TO OUR SOLAR SYSTEM

Our solar system is a collection of celestial objects that orbit around the Sun, which is a star located at the center. It is estimated that our solar system is about 4.6 billion years old and it consists of eight planets, five dwarf planets, comets, asteroids, and other celestial objects.

The eight planets in our solar system are categorized into two groups: the inner planets and the outer planets. The inner planets, including Mercury, Venus, Earth, and Mars, are commonly referred to as terrestrial planets since they have a solid rocky surface. The outer planets, including Jupiter, Saturn, Uranus, and Neptune, are referred to as giant planets since they are much larger than terrestrial planets and have a gaseous surface with no solid surface.

Apart from the planets, our solar system also has five recognized dwarf planets: Pluto, Ceres, Haumea, Makemake, and Eris. These dwarf planets do not have a clear path in their orbit and are much smaller than the eight major planets.

The region between Mars and Jupiter contains a vast number of asteroids, ranging in size from small rocks to objects several hundred kilometers in diameter. It is also home to the asteroid belt, a region of the solar system that is densely populated with asteroids and other minor planetary bodies.

Comets are also part of our solar system. They are icy objects that originate from the Oort Cloud and Kuiper Belt, areas far beyond the orbit of Pluto. Comets have highly elliptical orbits and can be visible from Earth as bright, fuzzy objects with a long tail.

Our solar system has an immense influence on Earth and its inhabitants. The Sun provides light and heat, making life possible on the planet. The Moon, our nearest neighbor in space, causes tides and is responsible for regulating Earth's rotation on its axis. The other planets also play a role in maintaining the stability of the solar system.

In conclusion, our solar system is a complex and fascinating area of study. It contains a wide variety of celestial objects, each with its own unique characteristics. The exploration of our solar system continues to provide new insights into the origin and evolution of our corner of the universe.

THE SUN: OUR SOLAR SYSTEM'S STAR

The Sun is the star at the center of our Solar System, and without it, life on Earth would not exist. It is a yellow dwarf star, meaning it is a main-sequence star in a stable state of nuclear fusion, converting hydrogen into helium at its core. The Sun's gravitational pull holds the planets in orbit around it, and its magnetic field and radiation have a significant impact on the physics and chemistry of the planets and other bodies in the Solar System.

The Sun has a diameter of about 1.39 million kilometers, which is about 109 times that of Earth. It is also much more massive, with a mass of approximately 2×10^{30} kilograms, or about 333,000 times the mass of Earth. The Sun's surface temperature is about 5,500 degrees Celsius, but its core temperature is a staggering 15 million degrees Celsius.

The Sun is made up of mostly hydrogen (about 71%) and helium (about 27%) with small amounts of other elements such as oxygen, carbon, neon, and iron. The layers of the Sun can be divided into the core, the radiative zone, and the convection zone, which are responsible for the nuclear fusion reactions that produce the Sun's energy.

The Sun's energy is produced by a process called nuclear fusion, which occurs when the temperature and pressure at the core of the Sun are high enough to force atomic nuclei to merge into heavier elements, releasing an immense amount of energy in the process. This energy manifests itself in the form of light and heat that radiates from the Sun and provides the energy that drives photosynthesis in plants and therefore sustains life on Earth.

However, the Sun can also be a source of danger. Its activity, such as solar flares and coronal mass ejections, can disrupt communication systems and even pose a threat to human health and technology in space. The Sun's effects on the Earth's climate and weather patterns have been the subject of much scientific study and research.

In conclusion, the Sun is the foundation of our Solar System and provides the energy necessary for life to exist on Earth. Its immense size and power continue to fascinate and influence human understanding and exploration of space.

INNER PLANETS: MERCURY, VENUS, EARTH, MARS

The inner planets of our solar system are Mercury, Venus, Earth, and Mars. They are called the inner planets because they are the four closest planets to the Sun.

Mercury is the smallest planet in the solar system, and it is also the closest to the Sun. The planet has a very thin atmosphere and no moons. Mercury's surface is rocky, and it is heavily cratered. The planet's surface is hot enough to melt lead during the day, but at night the temperature can drop hundreds of degrees below freezing.

Venus is the second planet from the Sun and is sometimes referred to as Earth's sister planet because they share a similar size and composition. However, Venus has a thick, toxic atmosphere that traps in heat, making it the hottest planet in the solar system. The planet's surface is very volcanic, and it has no moons.

Earth is the third planet from the Sun and is the only planet in the solar system known to support life. Earth's atmosphere contains oxygen, which is crucial for life as we know it. The planet has a relatively large moon, which helps stabilize its orbit and contributes to Earth's tides.

Mars is the fourth planet from the Sun and is often called the Red Planet because of its reddish appearance. The planet has a thin atmosphere, and its surface features include mountains, valleys, canyons, and vast plains. Mars has two small, irregularly shaped moons.

All four inner planets are rocky, or terrestrial, planets, meaning they have a solid surface. They all have relatively small sizes and masses compared to the outer, or gas giant, planets. They also all have shorter orbital periods around the Sun, meaning they take less time to complete one orbit.

In conclusion, the inner planets of our solar system are the four closest planets to the Sun: Mercury, Venus, Earth, and Mars. They are all rocky, have thinner atmospheres than the outer planets, and have shorter orbital periods. Each of these four planets has its unique characteristics that make them fascinating to study and explore.

OUTER PLANETS: JUPITER, SATURN, URANUS, NEPTUNE

The outer planets of our Solar System, Jupiter, Saturn, Uranus, and Neptune, are also known as gas giants. Unlike the inner planets, they are composed mostly of hydrogen and helium and have a much larger size and mass compared to the latter.

Jupiter, the largest planet in our Solar System, is known for its Great Red Spot and has a total of 79 known moons. Saturn, the second-largest planet, is famous for its beautiful rings, which are not solid but are made of small ice particles, boulders, and dust. Uranus and Neptune both have rings, but they are fainter compared to Saturn's rings.

Uranus has a unique feature; its rotational axis is tilted at a 98-degree angle, meaning it practically rolls on its side around the Sun. Neptune is the farthest planet from the Sun and has the strongest winds of any planet in our Solar System, with recorded speeds exceeding 1,500 miles per hour.

Just like the inner planets, these gas giants have their own moons, some of which are larger than the smallest planet in our Solar System, Pluto. Jupiter's moon Ganymede is the largest moon in our Solar System, even larger than the planet Mercury. Europa and Enceladus are two other moons that have garnered significant attention from scientists, as they are believed to have subsurface oceans that could harbor life.

The discovery of these outer planets changed our understanding of the Solar System and led to new questions about its formation. These gas giants played a critical role in shaping the Solar System's structure by pushing the smaller bodies away from the inner planets and towards the outer regions and beyond.

Despite being harder to study than the inner planets, these gas giants continue to fascinate scientists and inspire new discoveries. The exploration of these outer planets has led to the discovery of new moons and rings, changes in weather patterns and auroras, and intriguing mysteries that still need to be solved.

DWARF PLANETS: PLUTO, CERES, ERIS

Dwarf planets are celestial objects in our solar system that have the necessary features that define a planet but do not have enough gravitational pull to have cleared its orbit of other debris. Unlike traditional planets, dwarf planets are much smaller in size and are typically found in the Kuiper Belt region of our solar system.

The International Astronomical Union (IAU) has currently recognized five bodies as dwarf planets within our solar system: Pluto, Ceres, Eris, Haumea, and Makemake. Among the recognized dwarf planets, Pluto is the most well-known.

Pluto was first discovered in 1930 and classified as the ninth planet in our solar system. But in 2006, the IAU reclassified Pluto as a dwarf planet because it did not meet the newly defined criteria for a planet. Pluto has an irregular orbit and is accompanied by its five moons, including Charon, its largest moon.

Ceres, on the other hand, was discovered in 1801 and is the largest object in the astcroid belt located between Mars and Jupiter. It was initially classified as an asteroid but was reclassified as a dwarf planet in 2006. Ceres is the only dwarf planet located in the inner region of our solar system and is the closest dwarf planet to Earth.

Eris is another dwarf planet located in the Kuiper Belt region and was discovered in 2005. It was initially thought to be larger than Pluto, but its size has been revised after further observations. Eris has only one known moon, Dysnomia, and takes around 558 years to complete one orbit around the sun.

The significance of dwarf planets lies in understanding the

formation and evolution of our solar system. By studying their properties, scientists can gain insights into the initial conditions of our solar system and the processes that led to the formation of planets and other celestial objects.

In recent years, several missions have been sent to explore certain dwarf planets, including Ceres, with NASA's Dawn mission, which arrived at Ceres in 2015 and sent back images revealing the presence of water ice on the dwarf planet. The upcoming Lucy mission by NASA is expected to visit several asteroids, including some near Jupiter, which are regarded as protoplanets and are significant in our understanding of the early history of our solar system.

THE ASTEROID BELT

The Asteroid Belt, located between the orbits of Mars and Jupiter, is a region of the solar system where thousands of asteroids, ranging in size from hundreds of kilometers to just a few meters, orbit the sun. The belt is approximately 300 million kilometers from the sun, and its width ranges from 1.7 to 4.2 astronomical units (AU).

Scientists believe that the Asteroid Belt formed as a result of the gravitational influence of Jupiter, which prevented the material in this region from coalescing into larger bodies, ultimately resulting in the formation of numerous asteroids. The belt contains millions of asteroids, and the total mass of these objects is estimated to be about 4% of that of the moon.

The first asteroid, Ceres, was discovered by Italian astronomer Giuseppe Piazzi in 1801. Since then, several other asteroids have been discovered in the Asteroid Belt, with more discoveries being made every year. Astronomers have identified the largest members of the belt, some of which are over 500 kilometers in diameter, while many others are much smaller.

Exploration of the Asteroid Belt began in the early 1970s when NASA's Mariner 9 spacecraft conducted a flyby of the asteroid Eros. Since then, several other missions have been sent to asteroids in the region, and in 2001, NASA's NEAR Shoemaker spacecraft even landed on the surface of the asteroid Eros.

Studies of the Asteroid Belt have provided valuable insights into the formation and evolution of the solar system. The composition of the asteroids in the belt has been found to be quite diverse, and their study has helped astronomers understand the conditions prevalent in these regions during the early history of the solar system. The asteroid belt is also of significant scientific interest because it may contain valuable resources, such as water and metals, that could be useful for future space exploration and development.

The Asteroid Belt remains a fascinating area of study for astronomers, and continued exploration and observation of the region will undoubtedly yield many more discoveries and insights into the origins and evolution of our solar system.

COMETS AND METEOROIDS

Comets and meteoroids are two of the most interesting phenomena in our solar system. They are both remnants of the early solar system and provide valuable insights into its origins.

Comets are essentially cosmic snowballs made of ice, rock, and dust. When a comet comes closer to the sun, the ice starts to melt, and gas and dust are released into space, forming a tail that can be seen from Earth. The gas and dust that comets release are also the source of meteoroids.

Meteoroids are small, often rocky or metallic, pieces of debris that come from comets or asteroids. Some meteoroids are the size of a grain of sand, while others can be as big as a small boulder. When a meteoroid enters Earth's atmosphere, it vaporizes due to the extreme heat, creating a streak of light known as a meteor or shooting star. Some meteoroids are large enough to survive their descent through the atmosphere and hit Earth's surface as meteorites.

Comets and meteoroids are both important for understanding the early solar system because they are essentially "fossils" from that time period. Scientists can analyze the composition of comets and meteorites to learn about the conditions in the solar system when they formed. They can also study the orbits of comets and meteoroids to learn about the gravitational forces at work in the early solar system.

In addition to their scientific significance, comets and meteoroids also have cultural and historical significance. Ancient cultures often interpreted comets as omens of disaster, while meteorites have been revered as religious artifacts. Today, many people still enjoy watching meteor showers and tracking the movements of comets.

In conclusion, comets and meteoroids are fascinating and important celestial bodies in our solar system. They provide valuable insights into the origins of our solar system and are also culturally significant. By studying these objects, we can deepen our understanding of the universe and our place within it.

THE KUIPER BELT AND BEYOND

Beyond the orbit of Neptune lies a region called the Kuiper Belt, which is a disc-shaped area made up of icy bodies that range in size from small rocks to dwarf planets like Pluto. The Kuiper Belt is located in the outer part of our solar system and contains remnants from the early solar system.

The Kuiper Belt was first discovered by astronomer Gerard Kuiper in 1951, but it wasn't until the 1990s that the first objects within the Kuiper Belt were observed. The Kuiper Belt is estimated to contain about 100,000 objects larger than 100 km in diameter, and it's believed that many more smaller objects exist within the Belt as well.

One of the most famous objects within the Kuiper Belt is Pluto, which was considered a planet until 2006 when it was reclassified as a dwarf planet. Pluto is the largest known object within the Kuiper Belt, with a diameter of approximately 2,377 km. Pluto has a highly eccentric orbit that brings it closer to the Sun than Neptune at times but takes it as far away as 50 astronomical units at other times.

Beyond the Kuiper Belt lies the Oort Cloud, which is a region that contains comets and other icy bodies. The Oort Cloud is thought to extend out to a distance of 100,000 astronomical units from the Sun, or about 1.87 light-years.

The Oort Cloud is named after Dutch astronomer Jan Oort, who theorized its existence in the 1950s. The Oort Cloud is believed to be a vast, spherical cloud of icy objects that orbit the Sun at very great distances. Some of these objects occasionally come closer to the inner solar system, becoming visible as comets.

Currently, no missions have been sent to explore the Kuiper Belt or the Oort Cloud, but this area of the solar system continues to be an area of interest for researchers. It provides a window into the early history of our solar system, and studying the objects within these regions can help us understand the formation and evolution of our solar system, the Milky Way galaxy, and the universe itself.

THE ORIGIN OF OUR SOLAR SYSTEM

Our solar system comprises the Sun and all the objects that orbit it, including planets, moons, dwarf planets, asteroids, comets, and other small bodies. The origin of our solar system is an intriguing subject that has been studied for centuries by scientists and astronomers alike. There are various hypotheses and theories about how the solar system was formed, and here we will explore some of the most common ones.

The Nebular Hypothesis
The most widely accepted theory regarding the origin of the solar system is the nebular hypothesis. According to this theory, the solar system formed from a massive rotating cloud of gas and dust called the solar nebula. Under its own gravity, the solar nebula contracted into a spinning disk-like structure, with the Sun forming at the center. The leftover material in the disk coalesced into planets, moons, and other objects that eventually formed the solar system.

Planetesimal Hypothesis
Another theory related to the origin of our solar system is the planetesimal hypothesis. This theory proposes that instead of forming directly from gas and dust, small planetesimals were the building blocks of the planets. Planetesimals are small, solid objects that formed by accumulation of dust and ice in the outer parts of the solar system. These objects then collided and stuck together, forming increasingly larger planetesimals that led to the formation of planets.

Gravitational Instability Theory
A third theory that explains how the solar system was formed is

the gravitational instability theory. This theory suggests that the solar system formed from a large, cold, and dense cloud of gas and dust that collapsed under its own gravity. It is said that the gravitational instability generated enough energy to start nuclear fusion, giving rise to the Sun. The remaining gas and dust formed the planets and other objects in the solar system.

Regardless of the theory that one may find most compelling, the study of the origin of our solar system continues to be an important subject in astronomy. Understanding how our solar system formed has provided clues to the formation of other solar systems throughout the universe. By continuing to study the nature of our solar system, astronomers are continuing to expand our knowledge of the vast universe beyond our own small home.

THE FOUR TYPES OF PLANETARY BODIES

The planets in our solar system are divided into four types of planetary bodies based on their composition and features. These four types are Terrestrial planets, Jovian planets, Dwarf planets, and Exoplanets.

1. Terrestrial Planets:

Also known as the inner planets, the terrestrial planets are found within the inner region of the solar system. These planets are smaller in size and mostly made up of rock and metal. The four terrestrial planets in our solar system are Mercury, Venus, Earth, and Mars. The surface of these planets is solid, and they have a rocky crust.

2. Jovian Planets:

The Jovian planets are also known as gas giants. They are much larger than terrestrial planets and are mostly made up of gases such as hydrogen and helium. These planets do not have a solid surface and are typically composed of many layers of gas that get denser and denser as you move towards the core. The four Jovian planets in our solar system are Jupiter, Saturn, Uranus, and Neptune.

3. Dwarf Planets:

Dwarf planets are a type of small planetary body that shares characteristics with both planets and asteroids. They are not classified as planets because they have not cleared their orbit of other debris. Currently, there are five officially recognized dwarf planets in our solar system, which includes Pluto, Ceres, Eris,

Haumea, and Makemake.

4. Exoplanets:

Exoplanets are planets that exist outside of our solar system. They are of significant interest to astronomers because they provide insight into the diversity and composition of planets outside of our immediate solar system. To date, thousands of exoplanets have been identified. These planets can be classified into different categories based on their size, density, and composition.

In conclusion, the classification of planets into four types of planetary bodies is fundamental to help us understand their characteristics. These classifications help astronomers to identify patterns and develop theories of planet formation and evolution in our solar system and beyond.

444444444444

LUNAR EXPLORATION: THE MOON

The Moon is the Earth's only natural satellite, and the fifth largest moon in the solar system. It is also one of the most intriguing objects in the sky, and for centuries, people have been fascinated by its beauty and mystery. The Moon has been explored by humans and machines, and the data collected has provided us with a wealth of information about the formation of the solar system.

The first successful lunar mission was the Soviet Union's Luna 1, which flew by the Moon in 1959. Since then, many other countries have sent missions to explore the Moon, including the United States, Russia, China, and India. The primary goal of these missions is to collect data and samples that can help us understand more about the composition, structure, and geology of the Moon.

One of the most famous lunar missions was NASA's Apollo program, which landed astronauts on the Moon between 1969 and 1972. The program consisted of six manned missions, and each one provided valuable information about the Moon's surface, geology, and atmosphere. The astronauts also brought back rock samples that provided insights into the Moon's formation and the early history of the solar system.

The Moon's landscape is characterized by large impact craters, vast plains, mountains, and valleys. The surface of the Moon is also covered with a layer of fine dust called regolith, which was created by millions of years of meteoroid impacts. The Moon does not have an atmosphere, so there is no wind or weather erosion to change the landscape.

One of the most important discoveries made by lunar explorations is that the Moon's surface contains a variety of minerals, including iron, titanium, and aluminum. Some of these minerals could be used to make products that are essential to modern technology, such as computer chips and solar cells.

Despite the wealth of information we have about the Moon, there is still much to learn. Future lunar missions are being planned by various countries, including the United States and China. These missions will help us understand more about the processes that shaped the Moon and the early history of our solar system.

SPACE MISSIONS TO OUR SOLAR SYSTEM

Exploring our solar system has been a fascination for scientists and researchers for many decades. Over the years, countries all over the world have sent spacecraft and rovers to study and gather data about different planets, moons, and other celestial bodies within our solar system. These missions aim to discover more about our neighboring planets and their geological makeup, understand how the solar system was formed, and possibly find evidence of life beyond Earth. In this chapter, we will discuss some of the most notable space missions to our solar system.

One of the earliest space missions to our solar system was the Pioneer program launched by NASA in 1958. The Pioneer 10 and Pioneer 11 spacecraft were sent to explore Jupiter and Saturn, respectively. These missions provided valuable data about the composition and atmosphere of these planets.

Another milestone mission was the Voyager program launched in 1977, which sent two spacecraft, Voyager 1 and Voyager 2, to explore Jupiter, Saturn, Uranus, and Neptune. These missions provided detailed images and information about the planets, their moons, and their environments. Voyager 1 remains the farthest human-made object from Earth, and it entered interstellar space in 2012.

In 1997, NASA launched the Mars Pathfinder mission, which included a rover named Sojourner, making it the first spacecraft to successfully land on Mars. Since then, multiple missions have been sent to Mars, including the Mars Reconnaissance Orbiter, Mars Exploration Rovers, and the Mars Science Laboratory, which includes the Curiosity rover.

In 2006, NASA launched the New Horizons mission, which aimed to study Pluto and its moons. In 2015, the spacecraft made a close flyby, and the mission provided the first detailed images and measurements of the planet.

In addition to these missions, space agencies and organizations from different countries have also launched missions to study other parts of our solar system. The European Space Agency launched the Huygens probe in 2004, which landed on Saturn's largest moon, Titan. Japan's Aerospace Exploration Agency (JAXA) launched the Hayabusa mission in 2003, which landed on the asteroid Itokawa and returned samples to Earth.

In the future, there are more planned missions to our solar system, including the Europa Clipper, which aims to study Jupiter's moon, Europa, and the Mars 2020 mission, which includes the Perseverance rover that will study the Martian environment and look for signs of microbial life.

In conclusion, space missions to our solar system have played a significant role in our understanding of the planets, moons, asteroids, and comets within it. These missions have not only provided valuable data about our neighboring celestial bodies but also helped in advancing technology and contributed to our curiosity and desire to explore the unknown.

GRAVITY: ITS ROLE IN OUR SOLAR SYSTEM

Gravity is the force that holds our solar system together. It plays such a critical role in our understanding of the universe that it is impossible to explain the system's workings without talking about gravity. It is the main reason why the planets orbit the Sun, the moons orbit their planets, and the comets and asteroids remain trapped inside the solar system. In this chapter, we shall delve deeper into gravity and its pivotal role in our solar system.

Gravity is a force of attraction that exists between any two objects in the universe, with the strength of the force being directly proportional to their mass and inversely proportional to the distance between them. In our solar system, the Sun exerts the strongest gravitational force, given its massive size. The eight planets orbiting the Sun are in their respective positions due to the gravitational pull of the Sun.

The force of gravity is central to the formation of our solar system. The current prevailing theory suggests that our system formed from a giant cloud of dust and gas. Gravity pulled the cloud's materials together, and eventually, they began to spin and flatten into a disk shape. The center of the disk became the Sun, while the remaining material formed into the planets, moons, asteroids, and comets that we see in our solar system today.

Gravity plays a crucial role in keeping the planets in their assigned orbits around the Sun. The planets' orbits are slightly elliptical, which means that they are oval-shaped rather than circular. Therefore, their distance from the Sun varies, but the force of gravity keeps them from careening off into space or falling into the Sun.

Gravity also allows moons to orbit around their respective planets, and asteroids and comets to remain trapped in the asteroid belt and Kuiper belt, respectively. Without the force of gravity, these objects would fly off into space or move in a straight line.

In conclusion, gravity is the glue holding our solar system together. It is responsible for the planetary movements, and without it, the system would disintegrate. Gravity is one of the most fundamental forces in the universe, and understanding its role in our solar system helps us to comprehend its relevance in the universal scheme of things.

OUR SOLAR SYSTEM'S MAGNETIC FIELDS

Our solar system is not just a collection of planets, moons, asteroids, and comets. It is also an intricate web of magnetic fields produced by the sun and the planets. Many of these fields are interlinked and interact with each other affecting not only the space around them but also the space surrounding the Earth.

The sun's magnetic field, known as the heliospheric magnetic field, extends throughout the entire solar system by virtue of the solar wind, which carries charged particles towards the outer planets. The heliospheric magnetic field is important as it helps to protect the solar system from cosmic rays and galactic radiation. The magnetic field acts like a shield, absorbing much of the energy from these particles before they can reach and damage the planets within the system.

As for the Earth's magnetic field, it helps sustain life on our planet by blocking out harmful solar wind particles and cosmic radiation. The magnetic field of the Earth is generated by the swirling motion of molten iron in its core. This field is responsible for the auroras observed in the polar regions and can affect the operation of satellites and other electronic equipment.

Jupiter has the largest and most powerful magnetic field of all the planets in our solar system. Its magnetic field is more than ten times stronger than that of Earth and is responsible for its intense radiation belts. The magnetic field of Jupiter is also the reason why it is able to trap and hold on to its numerous moons.

Magnetic fields are also present around the other gas giants in our solar system, Saturn, Uranus, and Neptune. Each planet has a unique magnetic field that interacts with the solar wind and the other planets' magnetic fields. The magnetic field of Saturn, for example, is notable for its complex structure with magnetic lines that cross each other.

In addition to the magnetic fields produced by the planets, there are also magnetic fields present in asteroids and comets. Recent studies have shown that some asteroids have their own magnetic fields, generated by the motion of charged particles within the asteroid. These discoveries have provided new insights into the formation and evolution of these objects.

In conclusion, our solar system's magnetic fields play an important role in shaping the environment of space around us. They offer protection from high-energy particles and affect the behavior of cosmic rays and other forms of radiation. Studying magnetic fields is essential towards better understanding our solar system and the broader universe.

THE SUN'S INFLUENCE ON OUR SOLAR SYSTEM

The sun is the most important object in our solar system, as it's not only the brightest and hottest object, but it also has the greatest influence on the planets and other bodies that surround it. The sun's magnetic field is responsible for shaping our solar system, and it's the reason why we see various phenomena such as sunspots, solar flares, and coronal mass ejections.

One of the ways in which the sun influences our solar system is through its magnetic field. The sun's magnetic field is created by the movement of charged particles, and it's responsible for the formation of sunspots, which are dark spots that appear on the sun's surface. The magnetic field lines of the sun's magnetic field are twisted and contorted, which results in the buildup of energy that can be released in the form of solar flares and coronal mass ejections.

Solar flares are sudden and intense bursts of radiation that can cause disruptions in the ionosphere and affect communication systems on Earth. Coronal mass ejections are massive bursts of charged particles that are expelled from the sun's corona, and they can also have a serious impact on Earth. Coronal mass ejections can create geomagnetic storms that disrupt power grids and satellite systems.

The sun's magnetic field also influences the solar wind, which is a steady stream of charged particles that flows outwards from the sun. The solar wind interacts with the magnetic fields of the planets, creating complex structures such as the magnetotail and the magnetosphere.

Another way in which the sun influences our solar system is through its gravity. The sun's gravity is responsible for holding the planets in their orbits, and it also determines the shape of their orbits. The closer a planet is to the sun, the greater the pull of gravity, which means that it will orbit the sun faster than a planet that is farther away.

The sun's gravity also plays a role in the formation of our solar system. The solar nebula, which was a cloud of gas and dust that eventually formed our solar system, was held together by the gravitational attraction of the sun. As the nebula collapsed, it spun faster and faster, eventually forming a flattened disk. Within this disk, dust and gas clumped together to form the planets and other bodies in our solar system.

In conclusion, the sun's magnetic field and gravity have a significant impact on our solar system. The sun's magnetic field shapes the solar system and is responsible for the various phenomena that we see, while its gravity holds the planets in their orbits and played a key role in the formation of our solar system.

THE SIGNIFICANCE OF OUR SOLAR SYSTEM IN ASTRONOMY

The study of our solar system has played a crucial role in expanding our understanding of the universe. Our solar system, comprising of a central star, several planets, and an assortment of minor celestial bodies, offers us a unique opportunity to learn about the formation and evolution of planetary systems.

One of the most significant contributions of our solar system to astronomy has been in helping us understand how planets form. Scientists believe that our solar system formed from a massive cloud of gas and dust that collapsed under its gravity. The resulting compression caused the cloud to heat up and form the protostar that would become the sun. At the same time, the leftover gas and dust in the cloud began to clump together to form the planets.

The rocky inner planets, including Earth, and the gas giant planets, such as Jupiter and Saturn, all formed differently, based on their distance from the sun and other factors. The study of our solar system provides us with a unique perspective on the formation and evolution of planetary systems, as well as how and why they differ.

Our solar system also offers insight into the conditions that make life possible. As we explore the planets and moons in our solar system, we are searching for signs of life or environments that can support life as we know it. For instance, Mars, the fourth planet from our sun, has long been a focus of interest for astrobiologists. Researchers theorize that Mars may have once been home to liquid water, a necessary component for life as we know it.

Studying our solar system also allows us to better understand the impacts of natural phenomena on our planet. Through studying the impacts of comets and asteroids on other planets in our solar system, we can better prepare for potential impacts on Earth in the future. The field of planetary defense, which aims to protect our planet from potential asteroid and other celestial body impacts, is a direct result of our study of our solar system.

Finally, understanding our solar system enables us to more deeply explore the possibility of life beyond our planet. One of the most tantalizing prospects in the field of astrobiology is the idea of finding exoplanets beyond our solar system that may be capable of supporting life. However, our understanding of our own solar system is essential in this search, as it allows us to develop better tools and methods for detecting exoplanets and understanding their potential for supporting life.

In summary, our solar system is essential to astronomers and scientists seeking to understand the universe beyond our planet. It provides insight into the formation and evolution of planetary systems, the conditions that make life possible, and the potential threats faced by our planet. Its study will continue to be a crucial part of the exploration and understanding of our universe.

THE OORT CLOUD

The Oort Cloud is a hypothesized spherical shell of icy objects that is believed to be the source of many comets that pass through the inner solar system. It is located in the outermost fringes of the solar system and extends out to around 100,000 astronomical units (AU) from the Sun. The Oort Cloud is named after the Dutch astronomer Jan Oort, who first proposed its existence in 1950.

The Oort Cloud is made up of two distinct regions: the inner Oort Cloud and the outer Oort Cloud. The inner Oort Cloud is believed to contain objects that orbit the Sun at distances of up to 20,000 AU, while the outer Oort Cloud is believed to extend from 20,000 AU to 100,000 AU from the Sun. Some estimates suggest that the Oort Cloud could contain up to 2 trillion icy objects, with a combined mass that is several times greater than the mass of Earth.

Although the Oort Cloud has never been directly observed, its existence is supported by observations of comets that have highly elliptical orbits that bring them close to the Sun. These comets are thought to originate from the Oort Cloud, as their orbits cannot be explained by the gravitational influence of the planets in the solar system alone.

The Oort Cloud is believed to have formed from the remnants of the protoplanetary disk, the cloud of gas and dust from which the solar system formed, that was ejected to the outermost reaches of the solar system during its early history. The gravitational influence of nearby stars and galactic tides are thought to perturb the orbits of objects in the Oort Cloud, causing some to be ejected into interstellar space or to fall into the inner solar system as comets.

The Oort Cloud is a significant area of interest for astronomers and astrobiologists, as it may contain clues to the origins of the solar system and the processes that led to the formation of life on Earth. Future missions to the Oort Cloud could provide important insights into the early evolution of the solar system and the potential for life beyond our planet.

THE FUTURE OF OUR SOLAR SYSTEM

As we look towards the future, mankind's exploration of the solar system is just beginning. While we have already made significant strides in exploring and understanding our celestial neighborhood, there is still so much more to learn and discover. The future holds great potential for advances in both technology and our understanding of the universe.

One of the most promising possibilities is the potential for discovering life beyond planet Earth. As we continue to study planets and moons in our own solar system, such as Mars and Europa, we may find evidence of conditions that are ideal for supporting life. With advancements in technology, we may even discover extraterrestrial life forms that are completely different from anything we have seen on Earth.

In addition to exploring for life, exploring the solar system can also uncover many new resources that can be valuable to us. For example, we may discover new sources of water, minerals, and other materials that could be used for future space missions and even on Earth.

Another exciting possibility for the future of our solar system is the establishment of long-term human settlements on other planets, such as Mars. This goal has already been identified by NASA, and other private companies are investing in developing the technology and infrastructure necessary to make such settlements possible.

Of course, these advancements in space exploration and settlement require a great deal of preparation and planning. One of the biggest challenges is the development of more efficient and reliable space travel technology, which would reduce the cost and risk of long-term missions. In addition, we will need to continue studying these celestial bodies to better understand their environments, characteristics, and potential hazards.

As we look to the future, it is clear that exploration of our solar system and beyond will continue to be a major focus for scientific research and discovery. Our understanding of the universe is constantly evolving, and with each new discovery, we are one step closer to unraveling the mysteries of the universe.

SUMMARY AND CONCLUSION
OF OUR SOLAR SYSTEM

In conclusion, our solar system is an intricate and fascinating place. It consists of a star, eight planets, dwarf planets, moons, comets, asteroids, meteoroids, a Kuiper Belt, an Oort Cloud, and many more objects. Each of these celestial bodies has unique features and characteristics that continue to captivate our attention.

The Sun, as our solar system's star, provides the necessary energy for life on Earth and influences the entire solar system. The four inner planets of our solar system, Mercury, Venus, Earth, and Mars, are unique in their own way, with Earth being the only planet known to harbor life.

The four outer planets, Jupiter, Saturn, Uranus, and Neptune, are massive gas giants with diverse characteristics that scientists are still studying. The three dwarf planets, Pluto, Ceres, and Eris, also add to the complexity of our solar system.

The asteroid belt, located between Mars and Jupiter, is a region that contains many rocks and debris. Comets and meteoroids, on the other hand, come from the outer reaches of our solar system and are visible as streaks of light.

Beyond the Kuiper Belt lies the Oort Cloud, a vast sphere of icy objects. The Oort Cloud is believed to be the source of many long-period comets that enter our solar system from the outer reaches of the universe.

The origin of our solar system remains a mystery, but theories suggest it formed from a cloud of gas and dust, which eventually collapsed under its own gravity. The study of our solar system has led to numerous space missions to explore planets and moons, gather data, and further our understanding of our place in the universe.

Gravity plays a significant role in our solar system, holding objects in orbit and shaping planetary bodies. The magnetic fields of the Sun and planets also play a crucial role in the solar system's characteristics.

Our solar system's significance in astronomy cannot be overstated. It has inspired many scientists and astronomers to continue exploring, studying, and learning. The future of our solar system and its exploration is exciting, with many more discoveries yet to be made.

In conclusion, we have covered various aspects of our solar system, from its components to its origin and exploration. Our solar system is a magnificent and complex system that has fascinated people for centuries and will continue to do so. We eagerly anticipate the discoveries that lie ahead in our exploration of the solar system.